HI-TECH JOBS WITHOUT COLLEGE

BE A CYBERSECURITY SPECIALIST

by Cynthia Kennedy Henzel

BrightPoint Press
San Diego, CA

BrightPoint Press

an imprint of ReferencePoint Press, Inc.
Printed in the United States

For more information, contact:
BrightPoint Press
PO Box 27779
San Diego, CA 92198
www.BrightPointPress.com

LIBRARY OF CONGRESS CATALOGING-IN-PUBLICATION DATA

Name: Henzel, Cynthia Kennedy, author.
Title: Be a cybersecurity specialist / by Cynthia Kennedy Henzel.
Description: San Diego, CA: BrightPoint Press, 2025 | Series: Hi-tech jobs without college | Audience: Grades 7 to 9 | Includes bibliographical references and index.
Identifiers: ISBN: 9781678209384 (hardcover) | ISBN: 9781678209391 (eBook)
The complete Library of Congress record is available at www.loc.gov.

CONTENTS

AT A GLANCE

- Cybersecurity involves keeping computers, networks, and data stored on computers safe.
- Cybersecurity specialists assess computer systems, stop hackers, and fix problems that may affect computer systems.
- Cybersecurity threats are a major problem. They cost companies and governments billions of dollars each year.
- Cybersecurity is one of the fastest-growing job markets.
- People can get jobs in cybersecurity through technical training.
- Some cybersecurity specialists are penetration testers. They try to break into computer systems to look for areas that need to be fixed.

- Cybersecurity specialists monitor security software, install new software, and train employees in other departments to use computers safely.

- Auditors assess a company's computer system to determine whether the company needs to update its security.

- Businesses often hire people with recent cybersecurity training even if they don't have a college degree.

GUARDIAN OF THE CYBER WORLD

Anthony Messina is a cybersecurity specialist. The term *cyber* means having to do with computers. The cyber world includes technology such as smartphones. It also includes information stored on computer systems. Messina's job is to stop hackers. Hackers are people who use computers in unusual ways to gain access to information, usually for illegal purposes.

Sometimes companies use hackers to steal data from other companies. This is a crime known as corporate espionage.

Messina works at a large company. Each day, he logs into the company's security system. He looks for problems. One day, a hacker might try to get into the system and steal data from the company. Another day, people who work for the company might be the problem. They could accidentally create a security risk. For example, they might choose a password that is easy to guess. Messina writes reports on new issues. He also follows up on past cases.

When Messina discovers a problem, he works with the incident response team. This team looks at the damage. Then, they work together to fix the problem. Finally, Messina contacts anyone affected by the security **breach**.

Being a cybersecurity specialist can be tough. Messina said, "To me, the most challenging part of security analyst work is overcoming that initial fear of the unknown. When I first started analyzing alerts, there would be alerts that left me baffled. . . .

Companies rely on cybersecurity specialists to keep valuable information safe. That can put a lot of pressure on specialists.

Cybersecurity specialists protect the banking information of people who shop online.

It was like being on the bomb squad and not knowing which wire to cut first."[1] Staying calm under stress is an important skill for specialists.

WHAT IS A CYBERSECURITY SPECIALIST?

Cybersecurity specialists do a wide range of jobs. Some specialists test computer systems by trying to hack them. They want to find ways to break in. This helps them make the system more secure. Specialists may also train employees on how to use **networks** safely.

The field of cybersecurity is growing fast. Almost every system needs people to keep it safe. Society is relying more and more on computers. This means cybersecurity specialists will become more important in the future.

CHAPTER ONE

EXPLORING CYBERSECURITY

People who work in security protect property. They may protect people's homes or businesses from robbers. Or they might protect people from harm. Cybersecurity specialists protect the information people store in computers. Bank records and other important data are often stored on computers. This data can be valuable to hackers, who might try to steal it.

Servers are computer systems that store and provide access to data. Some cybersecurity specialists protect the data held on servers.

Cybersecurity specialists protect entire computer systems. These systems run major parts of people's daily lives. They run transportation networks in cities. Computer systems run hospitals and schools, too. Cybercrime threatens these systems. Ginni Rometty is a former CEO of the company IBM. She said, "[Cybercrime is the] greatest threat to every profession, every industry, every company in the world."[2]

SECURITY BREACHES

Computer programmers write code to tell computers what to do. They use programming languages to write code. They write the software or apps that people use on their devices. For example, a user clicks on an email. Their computer follows

Coding is not required for most entry-level cybersecurity jobs.

coded instructions to open it. Code must be accurate. A tiny mistake can prevent a program from running.

A mistake in the code can let a hacker in. Hackers might also trick an **authorized** user into giving away information. One method hackers use is called the Trojan horse. Its name comes from an ancient Greek story. In the story, people hide in a large wooden horse. They use it to sneak into a city. In cybersecurity, a Trojan horse is software that looks real. But the software is a hacking tool. For example, Trojan horses can be sent as a link in an email. An authorized user might click on the link. This allows the hacker to get into the system.

Inside a system, hackers can change the code. They can steal or destroy data. They can even shut down the whole system. Hackers may put malware

Cybersecurity specialists who work on a team need teamwork skills and communication skills.

on a computer. This is software meant to harm a system.

IT TAKES A TEAM

Cybersecurity specialists often work on a team. People have different jobs. Some make sure everyone in the company follows security rules. Others may **install** or update security software. These updates are sometimes small repairs to fix software. Specialists also **monitor** cyber systems for attacks. They use software made to find problems in systems. When attacks happen, cybersecurity specialists must work quickly to stop the hacker.

Cybersecurity specialists do not only wait for attacks. The cybersecurity team includes auditors. Auditors study systems and find weaknesses. They think of ways to improve systems to prevent attacks. Penetration testers are also part of the team. They try to break into systems. Testers tell a company if they find a weakness in software. These weaknesses could allow a hacker to get in.

FROM SCHOOLS TO GLOBAL NETWORKS

Cybersecurity specialists work for top industries. These include e-commerce, finance, health care, and manufacturing. They also work for governments. They keep utilities and transportation systems safe. They protect important data. Some specialists work for the military. They keep military communications

Stolen military data can impact the security of a country. This makes cybersecurity especially important for militaries.

and data safe. Stéphane Nappo is the head of global information security for a French banking company. He explained how quickly a hacker can damage a company. He said, “It takes 20 years to build a reputation and [a] few minutes of cyber-incident to ruin it.”[3]

Cybersecurity specialists work in schools, too. Students work on school computer networks. They do research.

They turn in their work. They communicate with teachers and other students. The US Cybersecurity and Infrastructure Security Agency (CISA) has found that K–12 schools are not ready for cyberattacks. This is because schools have little money to spend on security. They often spend only 2 percent of their technology budget on keeping systems safe.

During the 2022–2023 school year, the FBI warned schools about cyberattacks.

School Hack

In 2023, schools in Minneapolis, Minnesota, had three hundred thousand files stolen. The school refused to pay a $1 million ransom. The hackers then put the files online. They contained personal information, such as health records. Many students wanted this data kept private.

As schools become increasingly digital, more cybersecurity specialists will be needed to protect student data.

It said that ransomware attacks were increasing. Ransomware is a type of malware. It locks users out of a system or steals data. It changes the data so that regular users cannot read it. The hackers then demand money to unlock the data. This money is called a ransom. If the school does not pay, students' personal data may be put online. Five million students had been affected by cybercrime in schools by 2023.

CHAPTER TWO

TRAINING FOR THE JOB

Some students interested in cybersecurity go to college. They can earn a degree in cybersecurity information technology (IT). It may be a 2-year or 4-year degree. But only 31 percent of cybersecurity specialists earn a 4-year degree in cybersecurity. According to a 2023 report, 70 percent of employers prefer entry-level experience in the field over college degrees in cybersecurity.

A job interview is a cybersecurity candidate's chance to show that they are qualified for the role.

To do well in cybersecurity jobs, people need curiosity. They need strong communication skills. And they need basic knowledge of cybersecurity.

One way to get basic knowledge is to take certification courses. A certification course teaches skills for a particular job. For example, CompTIA's Security+ is a basic cybersecurity certification. Other certifications cover subjects such as

After the COVID-19 pandemic, online courses became a more common way to gain training and experience.

cyberlaw and cybercrime. One course trains people to become Certified **Ethical** Hackers. Earning a certification usually takes less than 2 years. Certifications make it easier to get a job in the field.

Some certification courses are **accredited** by the Accreditation Board for Engineering and Technology (ABET). ABET makes sure that classes in technology, science, and engineering meet high standards. Certifications can often be used for college credits. Fifty-one percent of cybersecurity specialists begin by earning a certification.

BOOT CAMPS

Boot camps are short training programs. They often take less than 6 months.

Some have a final project. Boot camps prepare students to take certification tests. Boot camp graduates often get beginner jobs in IT. These include consultants, analysts, technicians, or auditors.

Boot camps may work with a university. Others are hosted by a business, such as Microsoft. Some are in-person classes. Others are online. These allow students to work at their own speed. The University of Minnesota has an online cybersecurity boot camp. Google also has an online cybersecurity class.

Some boot camps include both in-person and online instruction. The boot camp given by the University of California, Irvine has online instruction. Students then work in labs for

Many colleges and universities have computer labs where students, including those studying cybersecurity, can use computers to learn.

hands-on training. The course takes 24 weeks to complete.

MILITARY TRAINING

The US Army trains people to become cybersecurity specialists. Army Cyber Operations Specialists defend the army's satellites and ships from cyberattacks. They also defend weapons systems. The US National Guard trains specialists as well.

Army Cyber Operations Specialists must be US citizens. They must be 17 to 34 years old. They have to score well on two parts of the Armed Services Vocational Aptitude Battery (ASVAB) test. One part is General Technical. It tests reading and math reasoning. The other is Skilled Technical. It tests general science and mechanical understanding. Cyber Operations Specialists need 10 weeks of basic training. They then need 45 weeks of Advanced Individual Training.

The military also hires civilians as part of the US Army Civilian Corps. The organization offers jobs in cybersecurity and IT. The Corps is a good place to get experience in cybersecurity.

MIDDLE AND HIGH SCHOOL

The National Initiative for Cybersecurity Careers and Studies (NICCS) is a US government program. It encourages students to become cybersecurity specialists. Cybersecurity jobs require a high school diploma. The NICCS suggests students take math, science, and

In 2023, demand for cybersecurity in the military was very high. Not enough people were applying to be cybersecurity specialists.

The earlier students discover their interest in cybersecurity, the more time they have to prepare for a career after high school.

computer science in high school. A career in cybersecurity requires less math than a career in computer science. But it is still good to take as much math as possible in high school.

In 2020, the US Department of Education started a program called CTE CyberNet. It helps educators learn about cybersecurity. Those educators can then teach students about cybersecurity. Robbie Harris is an assistant principal at a high school in San Antonio, Texas. Her school is part of the CTE CyberNet program. Harris said that the program has helped make cybersecurity an important subject in the city's educational system.

CTE CyberNet also encourages students to become cybersecurity specialists. High school students can take courses that lead to technology certifications. CTE CyberNet prepares students for internships. Internships give students hands-on experience. Students can meet people working in cybersecurity jobs. Plus, students may go to industry meetings. They get to hear about what is new in cybersecurity.

Students interested in tech careers can join activities such as CyberPatriot. CyberPatriot hosts one-week camps in the summer. Student Sydney Brown went to one of these camps. "Learning what's deeper into a computer, not just what works in a computer but being able to decode what's inside that computer, that was my biggest takeaway of the camp,"

Cybersecurity Apprenticeships

There are many cybersecurity apprenticeships. An apprentice learns about a job by working with a professional. Apprenticeships give on-the-job experience. In 2023, there were at least 733 cybersecurity apprenticeship programs. They were registered with the US Department of Labor.

WHAT QUALITIES MATTER FOR CYBERSECURITY PROFESSIONALS?

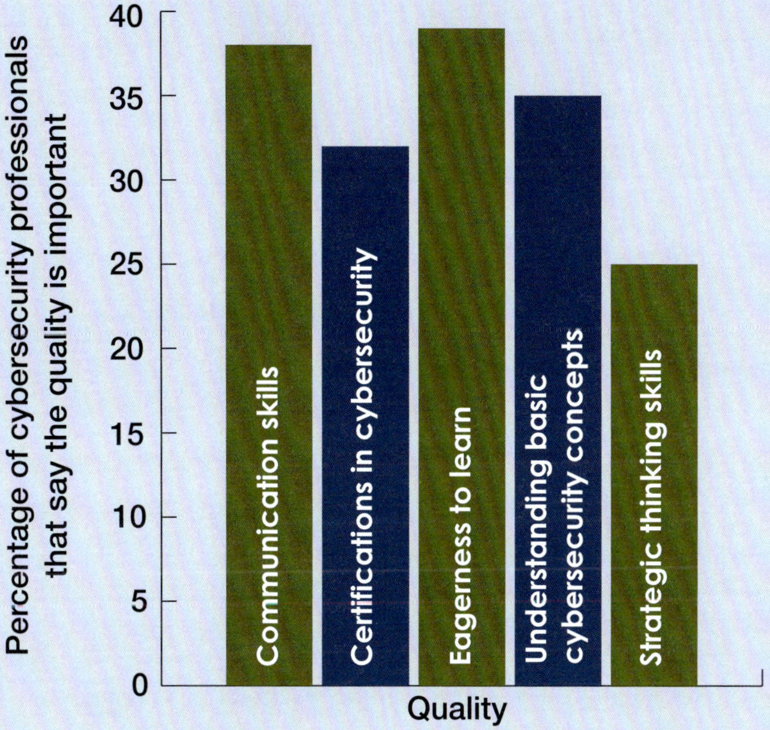

Source: "2023 International Information System Security Certification Consortium Cybersecurity Workforce Study," ISC2, October 31, 2023. www.isc2.org.

The bar graph above shows the qualities that were most often mentioned by cybersecurity professionals as important for people seeking a career in cybersecurity.

said Brown.[4] Other clubs, such as Future Business Leaders of America, provide cybersecurity events.

CHAPTER THREE

WORKING IN THE FIELD

Day-to-day work as a cybersecurity specialist depends on one's job. Some people work for a small company. They take care of all the company's cybersecurity needs. Others work for large companies. They are usually part of a security team. Each member plays their own role on the team. One person may monitor security software. Another person might repair secure systems. They then work together

Cybersecurity teams at a company are typically led by chief information security officers (CISOs).

as a team to solve problems when they happen.

Cybersecurity specialists do many things. They check businesses for possible problems. They may send fake emails to people working at the company. This is to test whether people follow security rules. If someone clicks on a fake link, the specialist knows they broke the rules. Specialists check that only approved devices use the system. Sometimes people work on home computers. Sometimes they use their own smartphone at work. These devices may cause a breach in the system.

Problems can happen. Specialists get to work quickly. They gather information about the problem. Then they fix it. After that, specialists talk to others in the company.

Cybersecurity specialists need to be able to help fellow employees follow cybersecurity guidelines.

They make sure the problem does not happen again.

Ian McEntire is a security analyst. He helps businesses improve their security. He works on security problems. His days are long. He said, "[S]ecurity incidents don't really care about when you're scheduled

Ethical hackers are also sometimes called white hat hackers.

to be working."[5] On one day, he explores possible security breaches. He meets with customers. He tells them how to improve security. He leads workshops about new security products.

Some specialists do penetration testing, or pen testing. Pen testers try to hack into computer systems. They are sometimes called ethical hackers. They test systems for weaknesses. Pen testers may use ethical hacking software such as Metasploit.

Pen testers report problems they find. They go to meetings to suggest new work methods or upgrades. Finally, they install software to fix problems. Devin Price is a pen tester. He said, "Being a penetration tester is largely a [non–people facing] role. Most of the time, you will be by yourself, alone, doing your job." He added, "What I have noticed is that the best [pen testers] are those who pay attention to aspects most people miss."[6]

AUDITORS

Security auditors **assess** the security of a company's network. Some companies hire auditors from outside the company. Other companies have auditors who work full-time. Auditors test a computer system for security problems. They then write a report about their findings. Auditors may help a company write safety rules for workers.

The government has laws to protect people's private data. Private data includes information about health care.

Many hospitals digitally store the private information of patients. Cybersecurity specialists protect this data.

Banking information is also private. Auditors make sure a system follows the law. Swathi West is an auditor. She works with health care companies. Her team has three members. She often works with other departments in the company. She said, "I love knowing that I'm helping to improve security not just for the organization but also for the people whose data they protect."[7]

Some cybersecurity specialists keep up with industry news by reading books and articles.

KEEPING UP

Any job in cybersecurity requires learning. There are always new things to learn. There is new technology to explore. Specialists spend time learning about their job. They talk to other cybersecurity specialists.

Specialists sometimes listen to podcasts by people in the industry. This helps them learn about new products or opportunities. They might also try out new software tools. All of this helps them do their jobs better.

In the past, most cybersecurity specialists were white men. This is

Valerie Thomas and NASA

Valerie Thomas was an early leader in cybersecurity. She led NASA's cybersecurity program from 1989 to 1993. Thomas helped NASA develop rules to protect its computers. She also worked to get women into cybersecurity careers. The US Congress awarded Thomas the Department of Commerce Gold Medal Award in 2017 for her work in cybersecurity.

Working from home became a more common option for cybersecurity specialists and other workers during the COVID-19 pandemic.

changing. In 2023, more men of color under 30 years old became specialists than white men. But there are still few women. In 2023, only 24 percent of specialists under 30 were women. Some organizations help women find jobs in cybersecurity. These include Women in Cybersecurity and Black

Girls Code. Minorities in Cybersecurity and Cyversity both provide opportunities for people of minority groups.

Job opportunities and pay depend on where one lives. Some states employ more cybersecurity specialists than other states. These include Virginia, Texas, and California. California, New York, and New Jersey have the highest salaries. About one-third of specialists work from home. Some may visit the office part of the time.

CHAPTER FOUR

LOOKING AHEAD

Companies need cybersecurity specialists who know new technologies. One of these technologies is cloud computing. Cloud computing refers to services that are provided over the internet. For example, people store pictures on the cloud. They may also use software stored on the cloud. Cloud computing security is a top skill that companies are looking for.

Fingerprint scanners can protect a device from being accessed by anyone except those whose fingerprints have been authorized for access.

esc
F1
F2
tab
Q
W
caps lock
A
shift
Z
fn
control
alt
option

Another new technology is biometrics. This technology identifies a person by using physical information. Fingerprints are biometrics. Face scans are biometrics, too. New types of biometrics can identify people by how they use a touch screen. Health care and e-commerce businesses need cybersecurity specialists to work in biometrics. The NICCS now offers certifications.

ARTIFICIAL INTELLIGENCE AND MACHINE LEARNING

Artificial intelligence (AI) is using computers to do things that usually require people. People give AI systems large amounts of data. This helps the AI solve problems. For example, a computer can scan millions of

Servers are sometimes used to store the large amounts of data scanned by AI.

pictures of dogs. It learns patterns. This helps it identify dogs in new pictures. A computer can scan millions of books. This helps computers write humanlike text.

AI lets computers make decisions. For example, a toy company might want to know how many Christmas toys to make. A computer looks at information about toys. It uses data that the company gives it.

It then predicts how many toys the company should make.

Machine learning (ML) is the practice of teaching computers to learn in a way that is like the way people learn. ML can allow computers to make decisions based on what they have learned. Many modern

Self-driving cars allow passengers to focus on other things as the car drives.

technologies and devices use ML. Self-driving cars are one example. Machine learning allows these cars to make safe driving decisions in new situations. But self-driving cars are not yet ready to drive by themselves. Human drivers are still required in certain situations.

AI and ML can be good and bad. New technology can make life easier. But criminals can also use it to do bad things. Internet crime can keep people from using new technology. Cybersecurity specialists must stop people who want to use AI and ML for bad purposes. In 2023, a report detailed trends in cybersecurity. The report found that AI and ML skills were in the top five most demanded skills.

THE INTERNET OF THINGS AND QUANTUM COMPUTING

The internet includes devices that share information with each other. These devices are called the Internet of Things (IoT). For example, the IoT includes chips placed in pets. The chips let owners track lost animals. The IoT also includes information sent from phones, cars, or homes. People's cars could tell their homes to turn on the heat before they arrive. Certain devices

Self-Driving Cars

The Internet of Things includes self-driving cars. In 2015, two researchers hacked a Jeep Cherokee. They gained control of the car's steering and brakes. More than a million cars were recalled to fix this problem.

Amazon Alexa, which follows spoken commands, is an example of an IoT service.

used by large cities are also part of the IoT. For example, one device might check a city's water supply. If the water is not safe, it uses the internet to sound an alarm. Or it may use AI to add cleaning chemicals. This fixes the problem.

IoT devices often have poor security. Many devices are easy to hack. More cybersecurity specialists are needed to protect these devices.

The switches in a computer associated with bits are called transistors. Modern transistors are too small to be seen by the naked eye.

The future of cybersecurity might involve quantum computing. Regular computers work with bits. These are the smallest units of information that a computer uses. A bit is in one of two states. The state depends on whether a switch inside the computer is turned on or off. Computers have millions or billions of these switches. Many bits together can form more complicated information. But bits limit how much data a computer can use. Quantum computers work with different kinds of bits. They allow the computers to do millions of calculations quickly. Quantum computers are still being worked on. One day, these computers may be much more powerful than regular computers.

Because of quantum computing, cybersecurity specialists must find new ways to protect computers. Regular computers often protect code by using difficult math problems. Hackers cannot solve the problems. But quantum computers might easily solve the math problems. A hacker with a quantum computer could break in. However, quantum computers might also help cybersecurity specialists. They could be a good tool to predict problems.

THE IMPORTANCE OF CYBERSECURITY IN THE FUTURE

Cybercrime costs people a lot of money. It may cost the world economy about $10.5 trillion each year by 2025. A majority of specialists reported that cybersecurity threats in 2023 were the most challenging of the past 5 years. Demand for specialists who can fight cybercrime is rising. Lenny Simon is a senior security engineer. He said, "Technology is always changing, and adversaries are getting more sophisticated. That's why there is so much growth."[8]

The demand for cybersecurity specialists makes it a promising career option for people without college degrees.

Being a cybersecurity specialist is an important career. The world is relying more and more on digital technology. That means the world will rely more and more on these specialists.

GLOSSARY

accredited
officially recognized as following specified standards

assess
to evaluate the abilities or qualities of an organization

authorized
allowed to do or use something

breach
a break or gap in a security system

ethical
having good morals or following the rules

install
to put software on a computer, usually after downloading it

monitor
to watch something closely

networks
systems of computers and technical devices connected together

SOURCE NOTES

INTRODUCTION: GUARDIAN OF THE CYBER WORLD

1. Quoted in Liz Simmons, "A Day in the Life of a Security Analyst," *Cyber Degrees*, December 8, 2022. www.cyberdegrees.org.

CHAPTER ONE: EXPLORING CYBERSECURITY

2. Quoted in Jessica Jury, "Cyber Security—What's the Big Deal?," *Victoria University*, n.d. www.vu.edu.au.

3. Quoted in "8 Tweetable Cybersecurity Quotes to Help You (and Your Business) Stay Safer," *Blue-Pencil*, n.d. www.blue-pencil.ca.

CHAPTER TWO: TRAINING FOR THE JOB

4. Quoted in Wes Tomlinson, "Teenagers Learn Programming, Operating Systems at Summer Camp," *Yahoo! News*, June 23, 2022. www.news.yahoo.com.

CHAPTER THREE: WORKING IN THE FIELD

5. Quoted in Rob Sobers, "What Working in Cybersecurity Is Really Like: A Day in the Life," *Data Security* (blog), *Varonis*, April 6, 2023. www.varonis.com.

6. Devin Price, "My First Year as a Pentester," *DPrice_TheBlogger* (blog), *Medium*, March 29, 2022. www.medium.com.

7. Quoted in James M. Tobin, "Day in the Life of a Security Auditor," *Cyber Degrees*, January 17, 2023. www.cyberdegrees.org.

CHAPTER FOUR: LOOKING AHEAD

8. Quoted in "A Day in the Life of a Cybersecurity Professional," *Information Technology* (blog), *Purdue Global*, November 30, 2020. www.purdueglobal.edu.

FOR FURTHER RESEARCH

BOOKS

Jennifer Connor-Smith, *The Wide World of Coding*. Minneapolis, MN: Lerner, 2020.

Erik Dafforn, *Information Technology (IT) Professionals*. Lanham, MD: Rowman & Littlefield, 2020.

Isabel Teitelbaum, *Be a Computer Support Specialist*. San Diego, CA: BrightPoint Press, 2025.

INTERNET SOURCES

"Cyber and IT Interns," *Cybersecurity and Infrastructure Security Agency (CISA)*, n.d. www.cisa.gov.

Aruna Pattam, "Artificial Intelligence, defined in simple terms," *HCLTech*, September 16, 2021. www.hcltech.com.

"What is Cybersecurity?," *Cybersecurity and Infrastructure Security Agency (CISA)*, February 1, 2021. www.cisa.gov.

WEBSITES

Air Force Association's CyberPatriot: The National Youth Cyber Education Program

www.uscyberpatriot.org

The CyberPatriot site has information on the program to encourage students to become cybertechnology specialists.

NASA Earthdata

www.earthdata.nasa.gov

This NASA site shows how the agency is using AI in Earth exploration.

National Initiative for Cybersecurity Careers and Studies

www.niccs.cisa.gov

The National Initiative for Cybersecurity Careers and Studies site provides information on cybersecurity careers for high school students. It also has information on internship and apprenticeship opportunities.

INDEX

IMAGE CREDITS

Cover: © Gorodenkoff/Shutterstock Images
5: © Gorodenkoff/Shutterstock Images
7: © Przemek Klos/Shutterstock Images
9: © PeopleImages.com-Yuri A/Shutterstock Images
10: © Natee Meepian/Shutterstock Images
13: © kubais/Shutterstock Images
15: © Gorodenkoff/Shutterstock Images
16: © Freedomz/Shutterstock Images
19: © Gorodenkoff/Shutterstock Images
21: © Monkey Business Images/Shutterstock Images
23: © PeopleImages.com-Yuri A/Shutterstock Images
24: © Chay_Tee/Shutterstock Images
27: © hxdbzxy/Shutterstock Images
29: © M_Agency/Shutterstock Images
30: © ESB Professional/Shutterstock Images
33: © Red Line Editorial
35: © fizkes/Shutterstock Images
37: © fizkes/Shutterstock Images
38: © Insta_Photos/Shutterstock Images
41: © Mr. Ashi. Sae Yang/Shutterstock Images
42: © ZoranOrcik/Shutterstock Images
44: © Olezzo/Shutterstock Images
47: © Demetrio Media/Shutterstock Images
49: © Wavebreak Media/Shutterstock Images
50: © Scharfsinn/Shutterstock Images
53: © Zapp2Photo/Shutterstock Images
54: © Scorpp/Shutterstock Images
57: © Gorodenkoff/Shutterstock Images

ABOUT THE AUTHOR

Cynthia Kennedy Henzel has a bachelor of science in social studies education and a master of science in geography. She has worked as a teacher-educator in many countries. She writes fiction and nonfiction books and develops educational materials for social studies, history, science, and English language learning students. She has written more than 100 books and 150 stories for young people.